Addendum to
Components for Evaluation of Direct-Reading Monitors for Gases and Vapors:

Hazard Detection in First Responder Environments

DEPARTMENT OF HEALTH AND HUMAN SERVICES
Centers for Disease Control and Prevention
National Institute for Occupational Safety and Health

This document is in the public domain and may be freely copied or reprinted.

DISCLAIMER

Mention of any company or product does not constitute endorsement by the National Institute for Occupational Safety and Health (NIOSH). In addition, citations to Web sites external to NIOSH do not constitute NIOSH endorsement of the sponsoring organizations or their programs or products. Furthermore, NIOSH is not responsible for the content of these Web sites. All Web addresses referenced in this document were accessible as of the publication date.

ORDERING INFORMATION

To receive documents or other information about
occupational safety and health topics, contact NIOSH at
Telephone: **1–800–CDC–INFO** (1–800–232–4636)
TTY: 1–888–232–6348
E-mail: cdcinfo@cdc.gov
or visit the NIOSH Web site at **www.cdc.gov/niosh**.
For a monthly update on news at NIOSH, subscribe to
NIOSH eNews by visiting **www.cdc.gov/niosh**/*eNews*.
DHHS (NIOSH) Publication No. 2012–163
July 2012

FOREWORD

The Occupational Safety and Health Act of 1970 (Public Law 91–596) assures, insofar as possible, safe and healthful working conditions for every working man and woman in the Nation. The act charges the National Institute for Occupational Safety and Health (NIOSH) with recommending occupational safety and health standards and describing exposure concentrations that are safe for various periods of employment, including but not limited to the concentrations at which no worker will suffer diminished health, functional capacity, or life expectancy as a result of his or her work experience.

Under that charge and by a 1974 contract, NIOSH and the Occupational Safety and Health Administration jointly undertook the evaluation of sampling and analytical methods for airborne contaminants to determine if current methods met the criterion to produce a result that fell within ±25% of the true concentration 95% of the time. In 1995, that protocol was revised.

The *Components for Evaluation of Direct-Reading Monitors for Gases and Vapors* expands the 1995 method development and evaluation experimental testing methods to direct-reading monitors for gases and vapors. It further refines the previous guidelines by applying the most recent research technology and giving additional experimental designs that more fully evaluate monitor performance.

This *Addendum* to the *Components* document expands the applicability of the *Components* by presenting methods to be used in evaluating direct-reading monitors for hazard detection in First Responder environments, including those related to incidents involving weapons of mass destruction (WMD). The *Addendum* contains a standardized test protocol and performance acceptance criteria for evaluating commercially available, direct-reading monitors in a style similar to the *Components* document.

John Howard, M.D.
Director
National Institute for Occupational Safety and Health
Centers for Disease Control and Prevention

ABSTRACT

The *Components for Evaluation of Direct-Reading Monitors for Gases and Vapors* (hereafter referred to as the *Components* document) [NIOSH 2012], presents methods to be used in evaluating direct-reading monitors* for use in workplace compliance determinations.

This *Addendum* to the *Components* document expands the applicability of the *Components* by presenting methods to be used in evaluating direct-reading monitors for hazard detection in First Responder environments, including those related to incidents involving weapons of mass destruction (WMD). The *Addendum* contains a standardized test protocol and performance acceptance criteria for evaluating commercially available, direct-reading monitors in a style similar to the *Components* document.

---

*Consistent with the *Components* document, the term *monitor* is used here to indicate a device for on-site measurement of contaminant levels for gases and vapors in air. The term *detector* refers to the component of the monitor that actually detects the contaminant.

CONTENTS

DISCLAIMER ii
ORDERING INFORMATION ii
FOREWORD iii
ABSTRACT iv
ABBREVIATIONS vii

Part I. Direct-Reading Monitor Background Information 1
    Introduction 1
    Definitions 4
        First Responder 4
        Ease of Use 4
        Consumables 4
        Maintenance Requirements 5
        Shelf Life 5
        Transportability 5
        Operational Limitations 5
        Environmental Conditions 5
        Class 1 Ensemble 5
        Life Cycle Analysis 5
        Direct-Reading Monitor 5
        Likert Scale 6
        Service Life 6

Part II. First Responder Environments 7

Part III. Suggested Components in Monitor Testing 9
    Physical Characteristics 9
        Documentation 9
        Transport Mode 9
        Ease of Decontamination 9
        Safety 10
    Operational Characteristics 10
        Operational Controls 10
        Alarms (Audible and Visual) 10
        Visual Display (Concentration Readout) 11
        Maintenance 11
    Performance Characteristics 12
        Environmental Effects 12
        Interferences 12

        Electromagnetic Interference 13
        Mechanical Stresses (Ruggedness) 13
        Decontamination Effects (Using Standard Procedure) 13
    Evaluation Panel for Ease-of-Use Characteristics 14
        Selection and Training of Panel Members 14
        Criteria and Procedures for Monitor Assessment 14
        Ranking Monitors for Ease of Use 16

References 17

References for Test Atmosphere Generation 18

Appendix A. First Responder Case Studies 19
    Train Accident in Baltimore Tunnel 19
    Terrorist Attack on World Trade Center 19

ABBREVIATIONS

AEL	authorized equipment list
ANSI	American National Standards Institute
ASTM	ASTM International
CBRNE	chemical, biological, radiological, nuclear, and explosives
°C	degree Celsius
dB	decibel
DHS	Department of Homeland Security
h	hour
in.	inch
IAB	InterAgency Board
IDLH	immediately dangerous to life or health
ISA	International Society of Automation
m	meter
mg	milligram
min	minute
NFPA	National Fire Protection Association
NIOSH	National Institute for Occupational Safety and Health
PPE	personal protective equipment
RKB	Responder Knowledge Base
SCBA	self-contained breathing apparatus
SEL	standardized equipment list
WMD	weapons of mass destruction

Part I. Direct-Reading Monitor Background Information

Introduction

Instrument capabilities and requirements must be defined for direct-reading monitors that are used as tools for First Responders in dealing with chemical, biological, radiological, nuclear, and explosives incidents. The definition of monitor requirements and the subsequent testing of individual monitors for product compliance will assist manufacturers in producing, and First Responders in selecting, commercially available instruments for measurement of personal exposure to hazardous substances when responding to emergency incidents.

Direct-reading monitors appear in Section 7 (Detection) of the Standardized Equipment List (SEL) developed by the InterAgency Board (IAB) [IAB 2012]. The SEL promotes interoperability and standardization across the responder community by offering a standard reference and a common set of terminology. The SEL has traditionally contained a list of generic equipment recommended by the IAB to local, State, and Federal government organizations in preparing for and responding to chemical, biological, radiological, nuclear, and explosive (CBRNE) events. The master, interactive version of the 2011 SEL, available at https://iab.gov/SELint.aspx, continues the transition to a broader "all-hazards" SEL, while maintaining an emphasis on CBRNE events. The interactive SEL provides mission-specific sublists designed to support critical mission areas. These sublists are compiled by subject matter experts who draw appropriate items from all 21 sections of the SEL as needed. Each sublist thus provides a "tailored SEL" for responders in a specific mission area. Another version of the SEL is available as part of the Responder Knowledge Base (RKB), at https://www.rkb.us. The RKB provides an SEL display with links to related standards, products, grants, and other equipment-related information, as well as an integrated display option that combines elements from the SEL and the Authorized Equipment List (AEL) produced by Department of Homeland Security (DHS).

Section 2 of the Homeland Security Act of 2002 (Public Law 107–296) defines emergency response providers as including "Federal, State, and local emergency public safety, law enforcement, emergency response, emergency medical (including hospital emergency facilities), and related personnel, agencies, and authorities." Homeland Security Presidential Directive/Hspd-8 (December 17, 2003) defines First Responders as "individuals who in the early stages of an incident are responsible for the protection and preservation of life, property, evidence, and the environment, including emergency response providers as defined in section 2 of the Homeland Security Act of 2002 (Public Law 107–296), as well as emergency management, public health, clinical care, public works, and other skilled support personnel (such as equipment operators) that provide immediate support services during prevention, response, and recovery operations."

Different categories of First Responders have different equipment needs and different hazard challenge conditions. Direct-reading monitors are employed by First Responders for exposure monitoring in: (1) the *crisis* phase, when quick decisions are required for the selection of personal protective equipment (PPE) for the response team, and (2) the *consequence* phase, when validated results are needed to allow others to go back into the site. The monitors are essential to locating the *hot zone*, to the positioning of decontamination stations, and to the control of personnel movement during the actual response to the incident.

The *Components for Evaluation of Direct-Reading Monitors for Gases and Vapors* (hereafter referred to as the *Components* document) [NIOSH 2012], presents methods to be used in evaluating direct-reading monitors for use in workplace exposure assessment determinations. Specifically, the objectives of the *Components* are:

(1) To provide guidance and procedures to estimate the precision, bias, and accuracy of a monitor. As for accuracy, the estimates include the single value that is the best descriptor of the accuracy, and a 90% confidence interval estimate. (Unless explicitly stated otherwise, all confidence interval estimates used in the *Components* are two-sided intervals.)

(2) For exposure assessment in routine processes, to provide guidance and procedures to evaluate a monitor relative to the 25% accuracy criterion (or one specified by the user) in terms of one of three mutually exclusive possible conclusions:

- A positive conclusion that there is 95% confidence that the monitor achieves the accuracy criterion. The monitor can be used for both compliance and range finding monitoring.
- A negative conclusion that there is 95% confidence that the monitor fails the accuracy criterion, i.e., that, at best, the method accuracy is worse than 25%. The monitor can only be used for range finding monitoring.
- An inconclusive finding that the monitor does or does not fulfill the accuracy criterion is inconclusive and that further research is required to resolve the question. The monitor can, at least, be used for range finding monitoring.

(3) To provide evaluation guidance for direct-reading monitors that need to demonstrate that an atmosphere is relatively *safe*. The most common usage of a *safe* determination involves situations where the monitor shows the concentration to be lower than a recognized occupational exposure limit.

This *Addendum* to the *Components* expands the applicability of the *Components* by presenting methods to be used in evaluating direct-reading monitors for hazard detection in First Responder environments, including those related to incidents involving weapons of mass destruction. The *Addendum* contains a standardized test protocol and performance acceptance criteria for evaluating commercially available, direct-reading monitors in the same style as the *Components* document.

Whereas monitor precision, bias, and accuracy are essential to workplace compliance determinations, these aspects are matched in importance by monitor ruggedness and ease of use for hazard detection in the typically harsher First Responder environments. Also, in the case of alarm-based monitors for First Responders, avoidance of false positives is more important than a high degree of measurement accuracy, because false positives trigger an intense response that is misdirected away from other potentially real hazards that should be addressed. However, it should be noted that if the monitor is accurate and specific, false positives should be greatly reduced.

Cost-effectiveness is another major issue that affects purchase priorities, whether the monitor is being used for workplace compliance assessment or for First Responder hazard detection. Besides purchase price, costs include expenditures for training, calibration supplies (and facilities), and periodic detector replacement that are typically more substantial, over the lifetime of the monitor, than the purchase price. The *Addendum* is written to ensure that First Responders are equipped with cost-effective direct-reading monitors that reliably meet their functional and tactical requirements, within the constraints of current technology.

This *Addendum* builds upon information presented in the *Components* document because the *Components* present information that is equally applicable to workplace compliance determination and to hazard assessment of First Responder environments. Part I of the *Components* presents definitions that apply to both situations, but additional definitions are necessary in describing the elements of the First Responder use of direct-reading monitors. Part II of the *Components* presents the principles of operation for the various types of commercially available direct-reading monitors, which apply to both types of application.

The application of direct-reading monitors to workplace compliance determination focuses on careful measurement of analyte concentrations in relation to hazard thresholds for worker exposure. The suggested components in monitor testing as presented in Part III of the *Components* reflect this focus. Besides examination of the ***physical characteristics*** of the monitor, the *Components* address the ***operational*** and ***performance characteristics*** as suggested components of monitor testing. ***Operational characteristics*** describe properties such as the ease of use, maintenance, calibration, and the results of ruggedness testing. ***Performance characteristics*** are determined by testing the monitor in atmospheres of known analyte concentration. It is important to note that the challenge conditions (e.g., operating temperature and humidity ranges) referred to in the *Components* correspond directly to those specified by the manufacturer of the specific monitor being tested.

The guidance described in Part III of this *Addendum* builds upon the test methods that are described in Part III of the *Components*. It adds challenge conditions that are independent of those specified by the manufacturer as an acceptable operating environment for the particular monitor. It also adds a spectrum of test methods for ***ease-of-use characteristics***. Because of the subjective nature of evaluating ease-of-use characteristics of a monitor, the use of an evaluation panel is recommended in making judgments about the how easy it is to operate and maintain the monitor in First Responder environments.

In most situations, the challenge conditions encountered by First Responders are much harsher than those encountered in the determination of workplace exposures. For example, First Responder conditions may be expected to include a condensing moisture environment, potential interferences from smoke and dust, poor lighting, and a high level of background noise. Although the *Addendum* considers challenge conditions that represent worst-case situations, it is recognized that there are many situations for which the actual challenge conditions are less intense.

In several cases, the *Components* references testing protocols for direct-reading monitors that have been published by NIOSH and other organizations, such as the International Society of Automation (ISA), the American National Standards Institute (ANSI), and ASTM International (ASTM). The availability of a well-developed testing protocol that is applicable to direct-reading monitors eliminates the need for preparing a protocol based on tests in the *Components* (or in the *Addendum*). Furthermore, it is stated that if a published standard protocol overlaps with that presented in the *Components*, the criteria associated with the more stringent testing should apply. These principles apply equally to this *Addendum*.

The *Components* (and similarly this *Addendum*) are designed for both the individual user and the manufacturer of direct-reading instrumentation for evaluation purposes. They can be used in part or in whole, depending on the need of the user. Manufacturers are encouraged to use the full range of this document for monitor evaluation. The *Components* and *Addendum* documents can also be used by consensus standard setting groups for preparation of specific standards for monitor performance.

The purpose of this *Addendum* is to further refine the established method evaluation guidelines for application to direct-reading monitor evaluation research and to provide additional experiments to more fully evaluate performance. An experimental design for the evaluation of direct-reading air monitors has been suggested. The experiments and definitions used in this document have been made compatible with those used by the ISA to the extent possible. If these experiments are not directly applicable to a monitor under study, then, a revised experimental design should be prepared that is appropriate to fully evaluate the monitor. The assistance of a statistician may be required for the preparation of this design.

Definitions

This section defines some terms that are used in the rest of the document. The reader is also directed to the definitions in Part I of the *Components*.

First Responder

First Responders are the earliest trained personnel on the scene when people's lives are in danger, such as from accidents, fires, and when hazardous substances are released. Traditionally, fire, police, and emergency medical service personnel have fit into this definition, but groups such as hospital workers, skilled support personnel, and utility workers should be considered as well, given the potential of their involvement in rapid response to mass disasters and exposure to risks. Also see the definition of First Responder by Homeland Security Presidential Directive/Hspd-8, as discussed in the Introduction.

Ease of Use

Ease of use refers to the extent to which the mobility and flexibility of the First Responder can be maintained while using the monitor. Ease of use also refers to ease of maintenance, the compatibility of the monitor with other equipment, and to the simplicity of monitor operation. Clarity of user documentation and intelligibility of visual or audible alarm signals are also included as ease-of-use characteristics.

Consumables

Consumables are those materials or components that are depleted or require periodic replacement through normal use of the instrument. Consumables include batteries, replaceable detectors, tubes, etc.

Maintenance Requirements

Maintenance requirements refer to the frequency of activities and necessary replacement parts required to keep a monitor in peak operational condition and ready for immediate deployment. Both preventive maintenance and repair/replacement activities are included.

Shelf Life

Shelf life is the length of time a monitor can be stored before it should be replaced with a new unit. Shelf life is defined according to the storage procedure and the required storage environmental conditions (e.g., temperature range). Typically the manufacturer specifies the shelf life as the warranty period, provided that the monitor is stored under the prescribed conditions.

Transportability

Transportability addresses the ease of moving the monitor to the desired sampling location, including any support equipment (e.g., hoses, battery packs). Transportability assumes the First Responder is wearing a self-contained breathing apparatus (SCBA) and a fully encapsulated chemical protective suit.

Operational Limitations

Operational limitations describe the range of operating conditions to which a monitor may be subjected without permanent impairment of operating characteristics.

Environmental Conditions

Environmental conditions refer to specified external conditions to which a monitor may be exposed during shipping, storage, handling, and operation. Temperature and humidity extremes, water exposure, and vibration/shock are typical environmental conditions.

Class 1 Ensemble

A Class 1 ensemble is a completely encapsulating, gas/vapor proof chemical resistant suit with a SCBA and chemically resistant, layered gloves and boots that are designed for the greatest level of skin, respiratory, and eye protection. First Responder protective gear designated Class 1 is designed for a chemical vapor or hazardous aerosol release and is defined by the National Fire Protection Association (NFPA) in *Protective Ensembles for Chemical/Biological Terrorism Incidents* [NFPA 2001]. There are other ensembles appropriate for First Responders, depending on the environment [NIOSH 2008].

Life Cycle Analysis

Life cycle analysis is the comprehensive examination of a product's environmental and economic effects throughout its lifetime, including new material extraction, transportation, manufacturing, use, and disposal.

Direct-Reading Monitor

A direct-reading monitor is a device that combines sampling and measurement of an analyte into one operation and offers real time or near real time exposure monitoring information. Direct-reading monitors are employed by First Responders for exposure monitoring in two situations: (1) the *crisis* phase, when quick results for the selection of PPE is required for the response team, and (2) the *consequence* phase when validated results are needed to allow others to go back into the site.

Likert Scale

The Likert scale is a measure of the extent to which a respondent agrees or disagrees with specified statements. It is derived from studies by industrial psychologist Rensis Likert [Likert 1932].

Service Life

The service life is defined as the length of time that the monitor can be operated before adjustments to the monitor need to be performed. Service life may be shortened by exposure to high levels of target contaminants that degrade components.

Part II. First Responder Environments

Performance specifications for direct-reading monitors used by First Responders must address risk-based standards for specified analytes (e.g., CO, H_2S, Cl_2, acid gases, radiation), while emphasizing monitor ruggedness and ease of use. These standards reflect the health effects of exposure and define hazard thresholds as a function of the exposure time. Analyte detection should extend from as low as 10 percent of the lowest occupational exposure level up to the immediately dangerous to life or health (IDLH) level and even beyond.

First Responder environments entail a high level of uncertainty about what may be present as hazardous substances. Moreover, the likely combination of multiple hazardous contaminants leads to a large probability of interferences that can lead to false positive or false negative monitor responses. The First Responder community is especially wary of false positives because they trigger much larger responses that may actually divert resources away from the most serious aspects of the hazards created by the incident. However, false negatives should also be considered is a similar way, since exposure to an unknown toxic hazard may have a greater impact on the response activities. This section addresses the harsh conditions of First Responder environments.

First Responder environments can be generated by: (1) biological incidents, (2) nuclear/radiological incidents, (3) incendiary incidents, (4) chemical incidents, and (5) explosive incidents. Examples of these five environments are presented in Table 1 for seven example threat scenarios. In each case, monitors must operate reliably when subjected to harsh conditions and in complex air environments.

Monitors used by First Responders may be exposed to toxic industrial chemicals (TIC) and to chemical warfare (CW), biological warfare (BW), and other hazardous substances. For example, the smoke from municipal structural fires contains many organic compounds in complex mixtures [Golka and Weistenhöfer 2008].

First Responders who arrive at the scene of an incident (whether indoor or outdoor) must assess the nature and extent of the incident. Based on available information, they may be able to estimate the primary and potential secondary emissions. First Responders will approach the vicinity of the incident from the upwind side (if outdoors) to evaluate the release type and note any casualties, especially those resulting from possible exposure to unknown vapors or aerosols.

The release type will define the monitors required by the First Responders (e.g., a ruptured tank car labeled with *chlorine* calls for a chlorine or halide monitor). In combination with observed casualties and meteorological conditions, the monitors are needed for personal protection and to establish a manageable perimeter, control zones and areas, and entry/exit points for First Responders.

Two case studies that illustrate the hostile environmental encountered by First Responders are presented in Appendix A of this *Addendum*. These case studies offer insights as to the harsh challenge conditions that should be represented when testing direct-reading monitors for First Responders. The recommended challenge conditions for monitor testing are specified in the testing procedures presented in Part III of this document.

Table 1. Example First Responder environments*

Threat scenario	Environment				
	Biological	Nuclear/ Radiological	Incendiary	Chemical	Explosive
Gas/aerosol release in subway	X	X		X	
Train derailment		X	X	X	X
Dirty bomb		X			X
Chemical plant explosion			X	X	X
Chemical truck or tank car accident		X	X	X	X
Plane crash into plant (chemical or nuclear)		X	X	X	X
Explosion in theatre			X	X	X

*These seven example threat scenarios can be compared to the fifteen national planning scenarios developed by the Homeland Security Council, which include 12 terrorist attacks (incorporating chemical, biological, radiological, nuclear, explosive, and cyber attacks) and three natural disasters (an earthquake, a hurricane, and a pandemic influenza outbreak).

Part III. Suggested Components in Monitor Testing

This part will describe, in detail, the suggested requirements and tests used when evaluating a portable, direct-reading monitor for use in harsh conditions. It will divide the requirements and tests into physical, operational, and performance categories, providing details of the requirement or test, and how the results of the requirement or test should be interpreted.

For specific tests, the required number of repetitions and suggestions on how to best conduct the test are included. These tests should be conducted with the specific analyte(s) for which the monitor was designed, if possible. If this is not possible, then chemically or biologically appropriate surrogate analytes may be used.

If a standard by the ISA, ANSI, or ASTM applies to the monitor under evaluation and is more stringent than those in this *Addendum*, the more stringent criteria defined under that standard should apply to the evaluation of the monitor. If the suggested requirements included in this *Addendum* are more stringent, then the monitor testing should address both the ISA/ANSI/ASTM standard and the applicable section of the *Addendum* as well.

Tests that are indicative of intramonitor variability are described in Part III of the *Components*. If these tests are performed on more than one monitor of the same type, estimates of intermonitor variability can be computed (See Appendix C of the *Components*). This provides a more realistic estimate of how the user may expect the monitor to perform.

Physical Characteristics

Physical characteristics document such properties as the size, shape, weight, and detection method of the monitor. This includes documentation for instrument operation, maintenance, and training. This *Addendum* adds two physical characteristics (transport mode and ease of decontamination) to what is already addressed in the *Components* (descriptive information, physical information, portability, and design).

Documentation

Each monitor manufacturer should provide documentation on the operation, maintenance, and theory of operation for the specified monitor. Documentation should be provided in a durable hard copy for use in responding to incidents. It should provide an easy means to find the operating instructions for the monitor. If the monitor is designed to be operated by a technician, then the operational instructions should be clearly written with common problems discussed in lay terms. If the monitor is designed to allow user maintenance, those maintenance procedures should be clearly specified. Any necessary parts for maintenance should be listed.

Transport Mode

This refers to the method for carrying the monitor, for example, the method of attachment to the responder, taking into account the level of personal protective equipment being worn.

Ease of Decontamination

This is related to the materials of construction and the structural design of the external shell of the monitor and its resistance to decontamination agents. Decontamination options include gas-phase chemicals, solutions, gels and foams, nanomaterials, electrostatic systems, and radiation. Typically, a smooth chemically resistant exterior surface with a minimum of seams aids in the decontamination process.

Safety

The manufacturer should provide instructions for the safe operation of the monitors. These should include any specific warnings about procedures or situations that may be hazardous to the operator or others in the general vicinity of an operating monitor. Safety precautions regarding calibration should be explicitly stated. The safety of the monitor in hazardous atmospheres should be stated, as well. If the monitor has been approved for use in flammable atmospheres, this should be indicated on the monitor and include the appropriate certifications. A common requirement for safety is compliance with the National Electrical Code definition of intrinsically safe [NFPA 2008].

Operational Characteristics

Operational characteristics describe properties such as the ease of use, maintenance, calibration, and the results of ruggedness testing. The suggested requirements given in the *Components* for operational characteristics also apply to First Responder hazard assessment.

When using direct-reading monitors for hazard assessment of First Responder environments, there are a number of ease-of-use operational characteristics that are as important as the more quantifiable monitor characteristics addressed in the *Components*. The importance of the ease-of-use characteristics stems from (a) the harshness of the First Responder environment, (b) the complicating effects of cumbersome personal protective equipment, and (c) the need to make quick decisions in assessing the nature of potential hazards caused by chemical, biological, radiological, nuclear and explosives incidents. The only ease-of-use characteristic appearing in the *Components* relates to defining the necessary sophistication of the operator.

Operational Controls

Operational controls should employ simplicity of design related to activator (button) size and location, and simplicity of function. The button, dial, or switch controls should be easy to use by First Responders with gloves and a face shield. A First Responder should be able to operate the unit without recourse to complex documentation. An evaluation panel of users would be employed to test the simplicity and functionality of operational controls against the best among commonly used technologies. The ease of use should be tested by asking each user (after a necessary training program) to execute a series of standard operating procedures with each monitor type. These procedures would include: (1) turning on/off, (2) zeroing, (3) internal calibration check (if available), (4) gas selection (if multi-gas), (5) reading the result, and (6) other functions. A special issue relates to using standard button controls when the user is wearing heavy gloves associated with a high protection level of personal equipment. As a practical guideline, the monitor should be operational within 10 min after arriving at the response site.

Alarms (Audible and Visual)

As compared to compliance environments, First Responder environments offer much greater challenges to the effectiveness of alarms. For audible/vibration alarms, sound level and sound distinctiveness are important. The effectiveness of visual alarms relates to brightness and observability (line of sight, color, flash sequence). Additional trouble indicators should include low battery and detector loss of function (e.g., break in circuitry).

The alarm properties listed under this performance area must be able to be seen or heard above background levels at the incident site. Moreover, an alarm signal must be distinguishable from other alarms inside any PPE that may be used, up to and including a Class 1 Ensemble. Test methods for proper alarm activation at hazardous concentration levels have already been developed. Other than sound level for an audible alarm, the effectiveness performance measures listed in this performance area also fall into the ease-of-use classification.

Audible signals will be evaluated using a standard test method for decibel strength at specified distances [ANSI 1983]. Once again, an evaluation panel of users would be employed to test the functionality and effectiveness of the alarms (and trouble lights) found on commonly used direct-reading monitors.

Visual Display (Concentration Readout)

For First Responder applications, the visual display should indicate analyte concentrations (standard units) and information on how the analyte concentrations compare to a specific occupational exposure limit or any predefined action level. The readability of the display is related to its size, color, and brightness.

The visual display should provide a digital readout of the analyte concentration in consistent units (e.g., parts per million), so that no units need to be converted. The display should also clearly indicate the hazard thresholds. The brightness and size of the concentration readout should be such that the numerical values of concentration can be readily distinguished over the range of likely First Responder challenge conditions. This would include a range of ambient lighting conditions and smoke concentrations that interfere with light transmission. The issue of face shields as related to the level of personal protection must also be dealt with.

None of the range of test procedures that have been developed in the past for evaluating portable direct-reading monitors provides anything beyond general acceptability criteria for visual display. There is also the difficulty of using a referee method for measuring a quantitative level of readability of the visual display. The evaluation of visual display is addressed in Section D, of this *Addendum*. A panel of users will assess display brightness and digit sizes available in direct-reading monitors.

Maintenance

Ease of maintenance provides a substantial savings in the life cycle of a portable monitor. Maintenance items include:
- Replaceable power supply
  - Common (flashlight batteries)
  - Optionally rechargeable
- Replaceable detector
- Calibration regime
  - Use of manufacture-supplied calibrant
  - Calibration flow system

The service life is defined as the length of time that the monitor can be operated before adjustments to the monitor need to be performed. Service life may be shortened by exposure to high levels of target contaminants that degrade components.

It is not recommended that maintenance be performed under difficult conditions or in potentially hazardous or corrosive environments. This may cause a failure to maintain the monitor properly and may lead to potential decontamination problems. It is assumed that monitor maintenance activities are performed before and after First Responder incidents.

Once again, this performance area falls into the category of ease-of-use. An evaluation panel of users would be employed to evaluate the comparative ease with which various maintenance functions on a particular monitor can be performed. For example, the evaluation panel will assess how easily batteries can be replaced, and the ease of cleaning a unit after exposure to dense smoke.

Performance Characteristics

Performance characteristics are quantifiable characteristics such as instrument sensitivity, specificity, response/recovery time, and response to interferences (chemical, electromagnetic, and environmental), mechanical stress, and decontamination procedures. Recommended methods for testing these characteristics are described below. These tests should be conducted with monitors that are representative of production monitors intended for emergency use.

A key element in the evaluation of a monitor is the generation of an atmosphere of known concentration to challenge the monitor. There are many different ways to generate an atmosphere and a discussion of such goes beyond the scope of the *Components*. A list of references on generation of test atmospheres is presented at the end of this document.

The accuracy with which a test atmosphere can be generated is key to a valid evaluation of a monitor. Often a second method must be used to verify the concentration generated. This method must be of known bias and precision. To facilitate the comparison of the reference method to the monitor results, appropriate statistical tests must be used to account for error in both the reference and test methods. See Appendix B of the *Components*.

Environmental Effects

The monitor should be exposed to extremes of temperature and relative humidity, as defined by the manufacturer, as well as intermediate temperature and humidity. When evaluating the monitor for use in routine exposure assessments, the extremes should correspond to manufacturer specifications as stated in the *Components*. However, when evaluating the monitor for hazard assessment by First Responders, the extremes are specified independent of manufacturer instructions.

Extreme environmental conditions have been defined by ANSI [ANSI 1990] for portable health physics instrumentation. For example, the recommended ranges of test conditions include:

- Temperature: –20 °C to 60 °C
- Humidity: 3% to 99% relative humidity, plus condensing atmosphere (fog)
- Rain (0.25 in./h)

Such conditions and associated test methods are also provided in military standard 810E [DOD 1989]. The monitor should be evaluated under these conditions.

After a standard test method is selected for each challenge condition parameter, the range of parameter variation must be adjusted to match the expected range for First Responders. These ranges may vary depending on the particular type and role of a First Responder. For example, in the *detect and confirm* role, firefighters are typically exposed to greater hazards than are border patrol officers.

Interferences

The effect of documented interferences on the operation of the monitor should be tested.

The measured magnitude of the interferences should agree with manufacturer specifications. If the monitor is to be used in a specific environment, then the effect of that environment on monitor performance should be checked.

Interferents found in First Responder environments are shown below:
- Particles
- Smoke
- Fog
- Dust
- Gases and vapors
  - Molecular structure similar to analyte
  - Detection properties similar to analyte
- Fuel vapors
- Fuel combustion emissions
- Aqueous Film Forming Foam (AFFF) used for firefighting
- Household chlorine bleach
- Engine exhaust

In all cases, an atmosphere of known concentration of an interferent must be generated, both in the presence and absence of the target contaminant. This is not a trivial matter unless a series of calibrants is available in a physical configuration such that the calibrant can be fed directly into the monitor without the need for any dilution. Alternatively, a dynamic test atmosphere can be generated by injecting pure target gas from a gas-tight syringe driven by a syringe pump at a specific rate (mg/min) into a larger gas stream (e.g., nitrogen or air) that is flowing through a duct. A static test atmosphere is typically generated in a gas-sampling bag. An airflow splitter is useful for detector comparisons and calibration experiments. More information on generation of test atmospheres and gas mixing can be found in the references at the end of this *Addendum*.

One of the complicating impacts of interferents is the potential for synergistic interaction between combinations of interferents that do not provide an additive effect. Tests may need to introduce interferent mixtures with specified level of the target compound (to distinguish false negatives), and interferent mixtures with no target compound (to distinguish false positives).

Electromagnetic Interference

Electromagnetic interference testing may be especially important for monitors used in the First Responder environment because of operation in close proximity to radio communications equipment. For further discussion and suggested testing, see the *Components* document Part III, Performance Characteristics, Electromagnetic Interference and Table 1.

Mechanical Stresses (Ruggedness)

Drop and vibration tests of the monitor would be conducted according to the specifications outlined in the ISA performance requirements [ISA 2010]. At a minimum, the monitor should be operational after a 1 m drop onto a hard surface. Failure of the monitor to operate after either the drop or vibration tests is indicative of failure of the monitor in the evaluation process. No further testing should be performed.

Decontamination Effects (Using Standard Procedure)

The ability to decontaminate the monitor after use in an aggressive environment with chemical or biological agents relates to materials of construction and the structural integrity of the exterior casing. The decontamination methods vary with the types and amounts of contaminants present on the monitor after use in a First Responder environment. Decontamination methods for testing should be applicable to the types of exposure anticipated for the monitor being considered.

As a practical guideline, the monitor should be capable of being purged and reset within 10 min after exposure to a heavily contaminated environment. This can be determined by measuring recovery response time, i.e., the time necessary to return to a background reading after exposing the monitor to a clean environment.

Evaluation Panel for Ease-of-Use Characteristics

A survey panel is formed to assess the ease-of-use operational characteristics of each monitor. This section describes the selection of the panel members and the design of the survey instrument to ensure that the evaluation is conducted in a balanced manner. Emphasis is placed on gathering semi-quantitative information in a time- and cost-effective manner so that a preferred direct-reading monitor can be selected for First Responder use. The proposed ease-of-use survey is expected to take less than four hours.

Practical experience shows that data collection from panel surveys may be confusing, so that the survey design and procedures presented below should be tested in a preliminary evaluation using a test panel to identify survey inconsistencies and problems. In particular, the responsibilities of the survey moderator should be clearly defined as the intermediary between the panel members and the designers and analysts of the survey.

Selection and Training of Panel Members

In this simple demonstration of an ease-of-use survey, the panel consists of four members with a mix of skills and experience suitable for the evaluation of four different direct-reading monitors. Each panel member must sign an informed consent document prior to participating. If relevant to the user audience of the monitors under test, factors of diversity, such as size, gender, culture, etc., should be addressed in the panel makeup.

Criteria and Procedures for Monitor Assessment

The panel members are informed by the test moderator of the purpose of the study, with the goal to identify the tested monitors most suitable to First Responder environments. The test moderator also describes the written survey questionnaire to the panel members and encourages them to add comments about each monitor. Panel members are instructed to read the operator's manual for each monitor before and during the time period that each monitor is evaluated.

The questionnaire provided to panel members should include the evaluation criteria for ease of use as listed below:

- Clarity of the operator's manual: Very clearly written and diagrammed, as compared to very poorly documented instructions.
- Transport and attachment for carrying by First Responder: Very easy to transport or attach to First Responder gear, as compared to very difficult to transport/attach.
- Characteristics of operational controls (e.g., buttons, switches): Very good tactile control of instrument function (even using gloves), as compared to very poor tactile control
- Characteristics of audible alarm signal: Very easy to distinguish particular alarm (even with high background noise), as compared to a very difficult to hear/inter-

pret alarm signal. Also indicated as decibels at 1 m distance.
- Visibility of monitor readout under obscured conditions and different types of lighting: Very easy to see monitor display (even with face mask and heavy smoke), as compared to very difficult to see/interpret monitor display.
- Ease of required instrument checkout, cleaning, and maintenance: Very little effort required, as compared to unusually difficult or lengthy procedures. Also indicated by time to complete regular maintenance, including monitor cleaning and battery replacement.
- Other ease-of-use conditions, as noted by the user.

The questionnaire will be designed according to a Likert scale that measures the extent to which a person agrees or disagrees with a given statement [Likert 1932]. Likert scale elements of 1 to 5 are recommended to assess ease of use:

(1) Very easy to use (alternatively, *very good performance*)
(2) Easy to use
(3) Not sure
(4) Difficult to use
(5) Very difficult to use

The questionnaire identifies the person conducting the survey, and is reproduced so that each panel member will respond to four questionnaires, one for each monitor type. In addition, soliciting user comments in responding to the questionnaires allows for unanticipated input important in the evaluation.

One monitor of each of the four types is selected and presented sequentially to the four panel members, in the order shown in Table 2. Panel members should evaluate monitors in random order to reduce any effects in monitor ranking from one time period to the next or from one monitor to the next. Panel member A sequentially evaluates monitors 1, 2, 3, and 4—in that order, while panel members B, C, and D evaluate monitors in different orders.

The four panel members concurrently evaluate the four monitors, in the random order indicated and for a time period to be determined. During each test, panel members are outfitted with typical protective gear and clothing, including face mask and double gloves, as specified. For purposes of evaluating the brightness and clarity of monitor display in a heavy smoke environment, welders' goggles (shade number ≈ 10), or the filters used for welders' goggles, are worn under the mask. A background noise is generated using sporadic amplified sound up to a sound level of 90 dB at 1 m.

Each panel member faces a different one of four walls in the test room so that the activities of other panel members are not observed. When a test period for a monitor is completed, a panel member moves to the table with the next monitor and documentation to be evaluated, using the sequences presented in Table 2. The panel members should be instructed not to reveal their scores or comments to others during the evaluation procedure.

After survey completion, members engage in a nonstructured panel discussion for a period of no longer than 30 min, so that the test moderator can record other useful comments. This discussion period is expected to bring out new evaluation criteria and to allow cross-comparisons, based on the users' perspective of having evaluated all four monitors.

Table 2. Sequence of monitor tests for each panel member

Panel member	Test sequence			
	First period monitor	Second period monitor	Third period monitor	Fourth period monitor
A	1	2	3	4
B	4	3	2	1
C	2	4	1	3
D	3	1	4	2

Ranking Monitors for Ease of Use

The questionnaire data should be encoded into a spreadsheet for evaluation. An independent reviewer should validate all data entry. Ordinal data (e.g., Likert scale measurements) should be analyzed by means of multiway contingency tables, and measures of agreement among panel members are estimated. Continuous data (e.g., alarm loudness, length of time needed for maintenance) should be summarized and the results expressed as mean ± standard deviations for each device. The monitor with the lowest Likert score will indicate the highest-ranking monitor, not considering comments and continuous data to the contrary. In addition, the comments sections of the questionnaire and the summary of the panel discussion will help reveal which monitors will perform best in a First Responder environment. The sex, age, education, and occupation of each panel member should be provided as part of the survey results. Finally, the ranking of the four monitors should be reported.

The reader is referred to the *Components* for discussions of Monitor Evaluation Data Reduction (Part III) and Evaluation Reporting and Documentation (Part III).

References

ANSI [1983]. American national standard: specification for sound level meters. New York: American National Standards Institute, Inc., ANSI S1.4-1983.

ANSI [1990]. American national standard: performance specifications for health physics instrumentation—portable instrumentation for use in extreme environmental conditions. New York: Institute of Electrical and Electronics Engineers, ANSI N42.17C-1989.

ANSI/ISA [2010]. ANSI/ISA-92.00.01-2010: Performance requirements for toxic gas detectors. Research Triangle Park, NC: American National Standards Institute, International Society of Automation.

Claudio, L [2001]. Environmental aftermath. Environ Health Perspect 109(11):A528–A536.

DOD [1989]. Military standard: environmental test methods and engineering guidelines. Washington, DC: U.S. Department of Defense, MIL-STD-810E [https://assist.daps.dla.mil/quicksearch/basic_profile.cfm?ident_number=35978]. Date accessed: July 2012.

Golka K, Weistenhöfer W [2008]. Fire fighters, combustion products, and urothelial cancer. J Toxicol Environ Health B Crit Rev 11(1):32–44.

Hartzell GE [1996]. Overview of combustion toxicology. Toxicol 115:7–23.

IAB [2012]. The InterAgency Board [https://iab.gov/default.aspx]. Date accessed: March 2012.

Likert R [1932]. A technique for the measurement of attitudes. Arch Psychol 22(140):1–55.

McClam E [2002]. Officials release WTC asbestos study. Associated Press, May 30.

NFPA [2001]. NFPA 1994: Standard on protective ensembles for chemical/biological terrorism incidents. 2001 ed. Quincy, MA: National Fire Protection Association.

NFPA [2008]. NFPA 70: national electrical code, 2008 ed. Quincy, Massachusetts: National Fire Protection Association, section 504.2.

NIOSH [2008]. Guidance on emergency responder personal protective equipment (PPE) for response to CBRN terrorism incidents. Pittsburgh, PA: U.S. Department of Health and Human Services, Centers for Disease Control and Prevention, National Institute for Occupational Safety and Health, DHHS (NIOSH) Publication No. 2008–132. [http://www.cdc.gov/niosh/docs/2008-132/pdfs/2008-132.pdf]

NIOSH [2012] Components for evaluation of direct-reading monitors for gases and vapors. Cincinnati, OH: U.S. Department of Health and Human Services, Centers for Disease Control and Prevention, National Institute for Occupational Safety and Health, DHHS (NIOSH) Publication No. 2012–162. [http://www.cdc.gov/niosh/docs/2012-162/]

References for Test Atmosphere Generation

Degn H, Lundsgaard JS [1980]. Dynamic gas mixing techniques. J Biochem Biophys Methods 3(4):233–242.

EPA [2007]. 40 CFR 53.22: Generation of test atmospheres [http://www.gpo.gov/fdsys/pkg/CFR-2007-title40-vol5/pdf/CFR-2007-title40-vol5-sec53-22.pdf]. Date accessed: March 2012.

EPA [2007]. 40 CFR 53.42: Generation of test atmospheres for wind tunnel tests [http://www.gpo.gov/fdsys/pkg/CFR-2007-title40-vol5/pdf/CFR-2007-title40-vol5-sec53-42.pdf]. Date accessed: March 2012.

Johnson DL, Fechter LD [1996]. Performance of an automatic feedback control vapor generation system during near-continuous inhalation exposures. Inhal Toxicol 8(4):423–431.

Johnson DL, Hagstrom EG, Fechter LD [1995]. Automatic feedback control of a vapor generation system using off-the-shelf components. Inhal Toxicol 7(9):1293–1303.

Nelson GO [1971]. Controlled test atmospheres: principles and techniques. Ann Arbor, MI: Ann Arbor Science.

Nelson GO [1992]. Gas mixtures: preparation and control. Boca Raton, FL: Lewis.

Shreeves K [1999]. The art of gas blending. Skin Diver, April; pp. 26–27 [http://www.skin-diver.com/departments/TotallyTech/Theartofgasblending.asp?theID=860].

Appendix A. First Responder Case Studies

Train Accident in Baltimore Tunnel

An actual example is useful to illustrate the harsh conditions encountered by First Responders and to offer insights that must be addressed when testing direct-reading monitors. A 60-car train accident produced a chemical spill complicated by fire in a tunnel underneath the former commercial center of Baltimore, MD. On July 18, 2001, eight tankers containing chemicals such as hydrochloric acid and tripropylene ruptured and released chemicals, fumes, and combustion aerosols into the tunnel air and the sewer system. The fire burned for five days following the accident. Magnifying the fire were creosote-soaked rail ties and contents of other railcars, including plywood, paper, soy oil, and scrap, that also were incinerated by the blaze in the over 100-year old, 1.7-mile long tunnel.

Extreme temperatures, fumes, toxic gases, and smoke complicated the duties of firefighters who ventured into the heat- and fume-filled darkness. The first firefighters on the scene entered the tunnel with 80 pounds of equipment, and picked their way through the blackness across rail ties, rails, and stone. Other firefighters entered nearby housing and told residents to close doors and windows for protection against smoke and fumes. Testing of gases continued around the clock at manholes in various locations above the tunnel. Inside the tunnel, testing was performed for hydrochloric acid, combustibles, oxygen, and carbon monoxide using several types of gas detection instruments.

Large movable fans were set up at one end of the tunnel to blow smoke and fumes away from the advancing firefighters. As the firefighters moved into the tunnel, the fans were moved forward to blow contaminants toward the opposite end of the tunnel. At the beginning of the fire, firefighters had open circuit SCBA with 30- and 60-min air supplies; however, these did not allow enough time to fight the fire in the hazardous environment. Firefighters were able to obtain from Dulles airport a total of 12 closed-circuit, positive-pressure SCBAs with four hours of breathing duration. The new equipment included an integrated breathing and air-cooling device, and an electronic pressure gauge for alarm pressure.

Emergency personnel at the Baltimore train wreck included firefighters, HazMat teams, the Coast Guard chemical strike force, and private HazMat contractors hired by the train company. While a few suffered heat exhaustion and smoke inhalation, not a single First Responder was seriously injured in bringing the disaster under control. In this example, monitors were required to perform well in extreme or harsh conditions, as compared with normal or standard test conditions.

Terrorist Attack on World Trade Center

The 9/11 terrorist attacks on New York's Twin Towers is another example of harsh environmental conditions that First Responders can encounter. The attack resulted in a hazardous brew of dust, soot, asbestos, and toxic combustion gases being released. Chrysotile asbestos was found in the dust and is a known inhalation health hazard. Dust also contained lead from lead-containing paint that was used to rust-proof steel beams in the Twin Towers. The article "Environmental Aftermath" cites asbestos, lead, and PCBs present in the dust created by the Twin Towers collapse [Claudio 2001].

The acts of terrorism that led to fires and collapse of the World Trade Center towers also produced noxious gases, including byproducts of combustion from the significant amount of solid building material and liquid fuel from the two aircraft. It is well known that toxic gases from structure fires typically include carbon monoxide and hydrogen cyanide. The gases and vapors listed in Table A–1 are released at nearly all fires from the combustion of construction materials and furnishings.

Following the World Trade Center attacks, the Environmental Protection Agency, the National Institute for Occupational Safety and Health, and the Occupational Safety and Health Administration monitored the environmental conditions at the World Trade Center and nearby areas. These agencies took air and bulk samples to test for silica, lead, carbon monoxide, noise, and numerous organic and inorganic compounds. OSHA alone took more than 6,000 samples. The dust and smoke from the numerous fires at Ground Zero included the following irritants and hazardous chemicals: silica, gypsum, fiberglass, paper, polyvinyl chlorides, pulverized concrete, benzene, and asbestos. However, Ken Wallingford, a researcher with the Centers for Disease Control and Prevention's National Institute for Occupational Safety and Health, has noted [McClam 2002], "The piece of data we don't have is what was the exposure to the folks who got caught in the dust cloud. That would've been a massive skin and inhalation exposure."

Table A–1. Toxic gases and vapors released at structure fires

Chemical	Source	Acute hazard
Acrolein	Carbonaceous materials	Irritant
Carbon monoxide	Carbonaceous materials	Asphyxiant
Carbon dioxide	Carbonaceous materials	Asphyxiant, respiratory stimulant
Formaldehyde	Carbonaceous materials	Irritant
Hydrogen cyanide	Nitrogenous materials	Asphyxiant

Halogen acids
  hydrogen bromide
  hydrogen chloride
  hydrogen fluoride
	Polymers and refrigerants	Irritant

Nitrogen oxides	Nitrogenous materials	Irritant
Isocyanates	Polyurethanes	Irritant

Source: Hartzell [1996].